SUPER-CAMOUFLAGED ANIMALS

BLACK RABBIT BOOKS | MEG GREVE

TABLE OF CONTENTS

1

The Chameleon

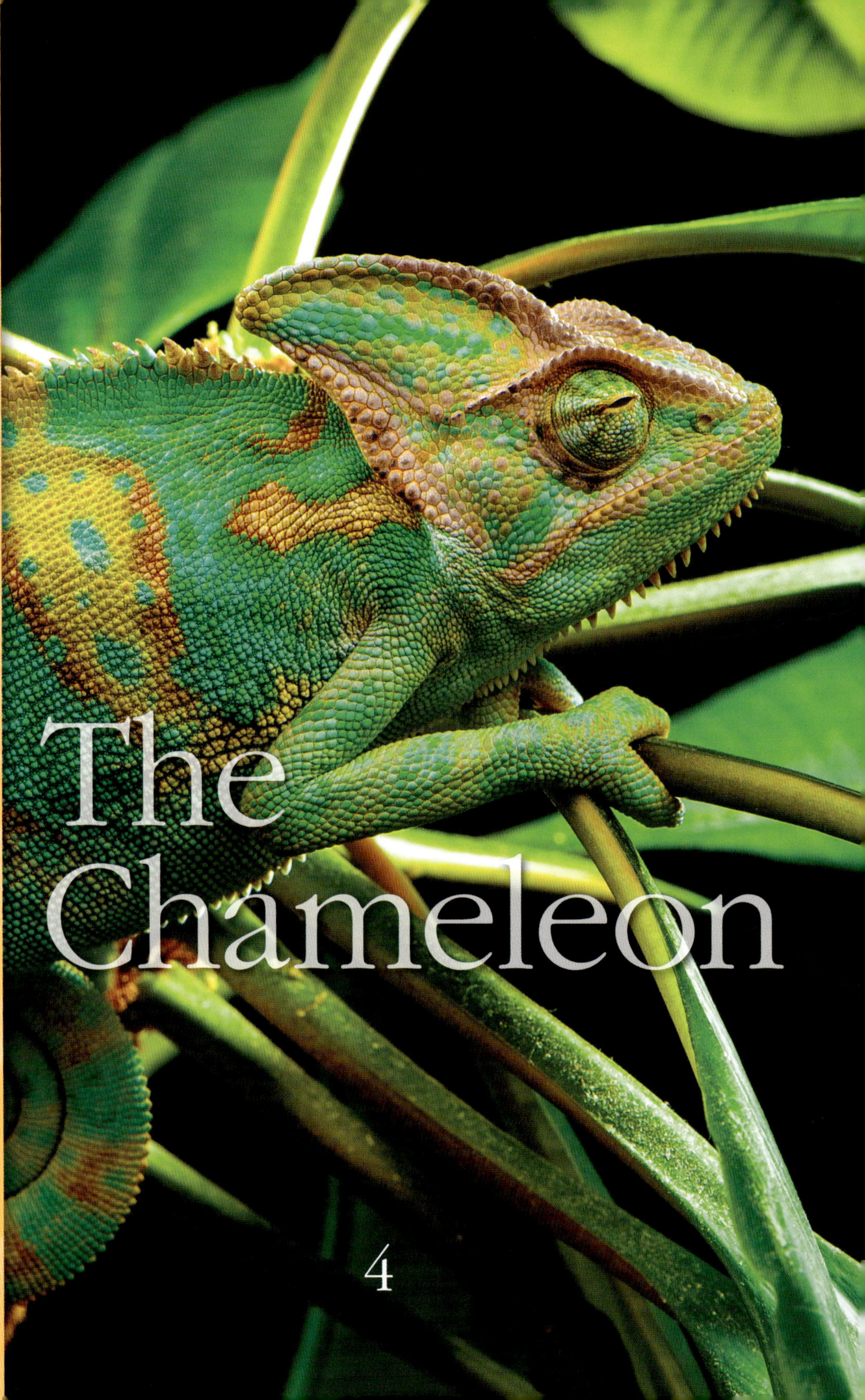

What color are chameleons? It depends! They can turn different colors based on their mood. A calm one might be light green. An angry one might turn red or yellow. It is ready to fight.

Chameleons have natural camouflage. They blend into what is around them. How? Their skin is clear. Underneath the skin, they have layers of special-colored cells. When these cells get bigger or smaller, the chameleon changes color.

Did You Know?

Chameleons turn colors to cool down too. A lighter color reflects the sun's heat. A darker color soaks up heat.

2

The Sea Dragon

Leafy sea dragons look like the seaweed. They hide in it. Leafy sea dragons can also change colors. They sway like plants as the water moves.

Leafy sea dragons live in the ocean near Australia. Their bodies are brown or yellow. The little round flaps on their bodies look like leaves. Weedy sea dragons live in the same area. They are red, yellow, and purple. Their fins are like feathers. They blend into the sea floor.

Did You Know?

Sea dragons are slow swimmers. They hide to avoid being eaten.

3

The Walking Stick

The walking stick freezes in place. This confuses its enemies. It looks exactly like a tree branch that predators would never eat. The walking stick can even sway back and forth when walking. Predators think it is a twig moving in the breeze.

Most walking sticks are brown or green. They can be as tiny as half an inch (1.3 centimeters) or as long as 13 inches (33 cm). Walking sticks live all over the world except in Antarctica.

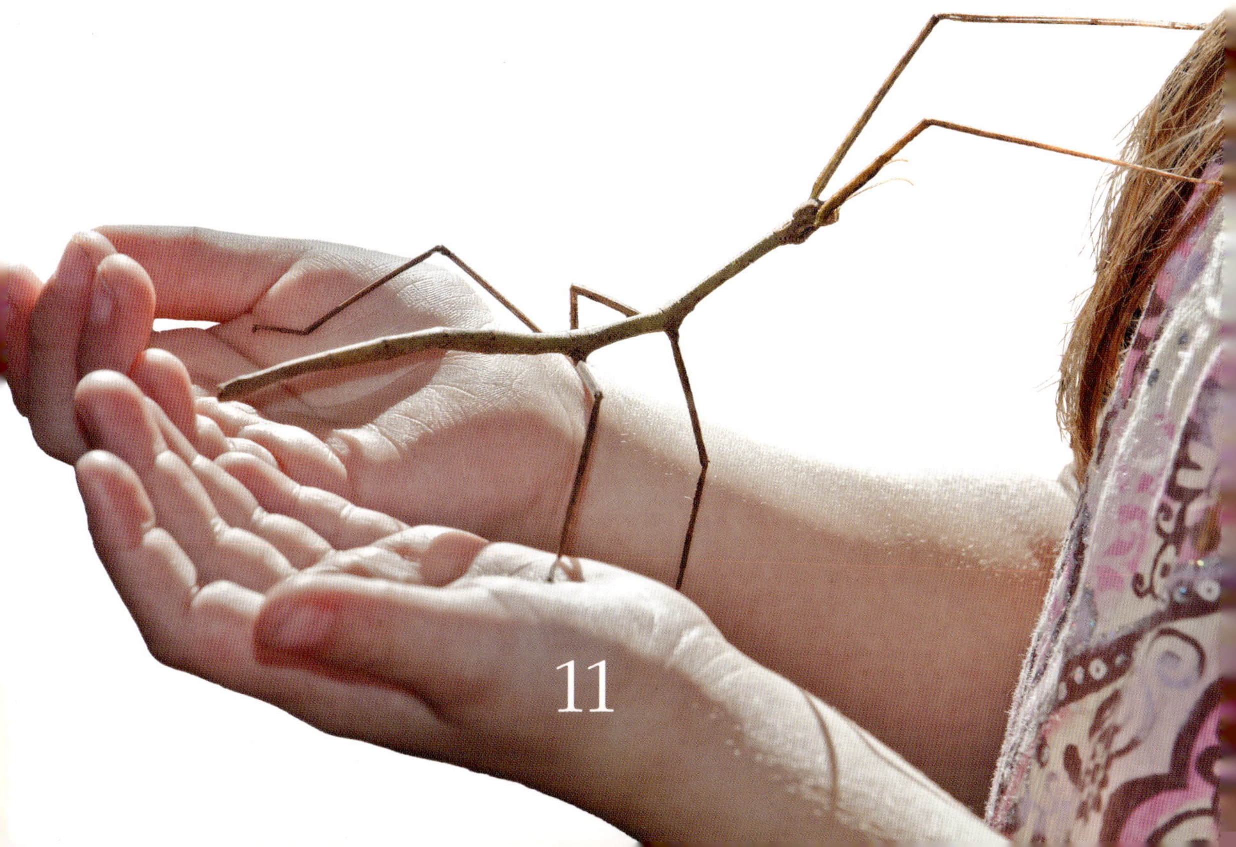

Did You Know?

When attacked, some walking sticks can remove their leg to get away. They can regrow a new leg.

4

The Coyote

Coyotes live in prairies, deserts, forests, and mountains. They have even moved into cities and neighborhoods. A coyote's fur matches where it lives. It can be gray to light brown.

Coyotes hunt small mammals, birds, and snakes. They use their senses of sight and smell to stalk the animals. Their thick fur makes them seem to disappear. Their prey does not stand a chance.

Coyotes live together in families. They like to hunt alone. Female coyotes have 4 to 7 pups each spring. Both parents feed and care for the pups.

Think About It
Coyotes have **adapted** to city life. Research what other animals have done the same.

5

The Snowshoe Hare

Snowshoe hares have two ways to protect themselves. First, they try to hide. If that doesn't work, they run away. They can race at 27 miles (43 kilometers) per hour!

Snowshoe hares' fur changes with the seasons. During the winter, they have white fur. They blend in with the snow. In the summer, the hares have reddish-brown fur. They blend in with dirt and rocks.

Predators have a hard time finding the hares. If they do, the hares can leap 10 feet (3 meters) in a single jump. Then they escape.

Did You Know? It takes about 10 weeks for the hare's fur to fully change color.

The Stonefish

6

Stonefish hide in shallow waters in the Indian and Pacific Oceans. They dig down into the mud or sand. Then they are hard to spot.

A stonefish waits for small fish and shrimp to swim by. It might stay in the same spot for many days. Once its prey comes close, the fish is quick to strike. It opens its mouth. Then it sucks up the animal like a vacuum.

The stonefish is also venomous. The venom is in spines on its back. Don't step on this fish!

Think About It

How might camouflage and venom help the stonefish?

MORE TO EXPLORE

WHERE IN THE WORLD?

SNOWSHOE HARES
North America

COYOTES
North America and Central America

WALKING STICKS
All over the world except Antarctica

CHAMELEONS
Africa, Asia, Europe, and Madagascar

SEA DRAGONS
Coast of Australia

STONEFISH
Indian and Pacific Oceans from the Red Sea to the Great Barrier Reef

MORE TO EXPLORE

FANTASTIC FACTS

Male chameleons are more colorful than the females.

Coyotes sing and howl at night.

Since stonefish are so well camouflaged, they cannot be easily seen. It is best for people to swim rather than walk in shallow coastal waters.

The only part of the snowshoe hare's fur that does not change is its ears. They have black tips on their ears all year!

Male sea dragons carry the eggs that the female lays.

Walking sticks lay up to 1,500 eggs!

MORE TO EXPLORE

COOL COMPARISONS

Spot the animal! Which one do you think has the best camouflage?

Chameleon

Sea Dragon

Walking Stick

Coyote

Snowshoe Hare

Stonefish

MORE TO EXPLORE

RESOURCES

Glossary

adapt (uh-DAPT) To change so it is easier to live in a different place.

camouflage (KA-muh-flazh) Coloring or covering that hides an animal in their surroundings.

mammal (MAM-uhl) A type of animal with hair or fur that drinks their mother's milk.

prairie (PRAYR-ee) A large, flat area with few trees that is covered in grass.

shallow (SHAL-oh) An area of water that is not very deep.

stalk (STAWK) To pursue prey in a sneaky way.

venomous (VEN-uh-mus) Capable of stinging or biting with venom.

Read More

Davidson, Rose. *Stripes and Spots*. Washington, DC: National Geographic Kids, 2023.

Smith, Penny. *Animal Hide and Seek*. New York: DK Publishing, 2023.

Index

TOP RANK is published by Black Rabbit Books, P.O. Box 227, Mankato, MN, 56002. • Top Rank is an imprint of Black Rabbit Books • Edited by Alissa Thielges • Designed by Danny Nanos • Photographs © Dreamstime/Elf0724, 8, Tsrapp, 6–7, 21; Shutterstock/Baishev, 4, 21, David W. Leindecker, 11, FotoRequest, 23, JayPierstorff, 13, Jim Cumming, 16, 17, 18, 21, John A. Anderson, 19, 21, 23, Kuttelvaserova Stuchelova, cover, 5, Lost_in_the_Midwest, 10, 21, 23, Luca Bertalli, 23, Mark Brandon, 12, mashe, 2, Mike Truchon, 15, 21, 23, Rich Carey, 20, Richard Seeley, 18, RobDemPhoto, 14, Roser Gari Perez, 9, 23 • Printed in China

Library of Congress Cataloging-in-Publication Data: Names: Greve, Meg, author. Title: Super-camouflaged animals / by Meg Greve. Description: Mankato, MN: Top Rank is an imprint of Black Rabbit Books, [2025] | Series: Super-incredible animals | Includes bibliographical references and index. | Audience: Ages 8–11 | Audience: Grades 2–3 | Identifiers: LCCN 2024017527 | ISBN 9781645823766 (library binding) | ISBN 9781645823988 (paperback) | ISBN 9781645824206 (ebook) | Subjects: LCSH: Camouflage (Biology)—Juvenile literature. | Adaptation (Biology)—Juvenile literature. | Classification: LCC QH546 .G74 2025 | DDC 578.4—dc23/eng/20240516 | LC record available at https://lccn.loc.gov/2024017527